YOUR KNOWLEDGE HAS VALUE

- We will publish your bachelor's and master's thesis, essays and papers

- Your own eBook and book - sold worldwide in all relevant shops

- Earn money with each sale

Upload your text at www.GRIN.com and publish for free

Bibliographic information published by the German National Library:

The German National Library lists this publication in the National Bibliography; detailed bibliographic data are available on the Internet at http://dnb.dnb.de .

This book is copyright material and must not be copied, reproduced, transferred, distributed, leased, licensed or publicly performed or used in any way except as specifically permitted in writing by the publishers, as allowed under the terms and conditions under which it was purchased or as strictly permitted by applicable copyright law. Any unauthorized distribution or use of this text may be a direct infringement of the author s and publisher s rights and those responsible may be liable in law accordingly.

Imprint:

Copyright © 2017 GRIN Verlag, Open Publishing GmbH
Print and binding: Books on Demand GmbH, Norderstedt Germany
ISBN: 9783668542082

This book at GRIN:

http://www.grin.com/en/e-book/376778/impacts-of-anthropogenic-climate-change-on-human-rights-a-look-at-the

Sonali Narang

Impacts of Anthropogenic Climate Change on Human Rights. A Look at the Treaths for India

GRIN Publishing

GRIN - Your knowledge has value

Since its foundation in 1998, GRIN has specialized in publishing academic texts by students, college teachers and other academics as e-book and printed book. The website www.grin.com is an ideal platform for presenting term papers, final papers, scientific essays, dissertations and specialist books.

Visit us on the internet:

http://www.grin.com/

http://www.facebook.com/grincom

http://www.twitter.com/grin_com

Impacts of Anthropogenic Climate Change on Human Rights

Dr. Sonali Narang, Assistant Professor, Department of Political Science, Arya College, Kurushetra University, Panipat, India.

Introduction..2

Climate change threatens the Right to Life in India 4

Climate Change threatens the Right to Adequate food in India 5

Climate Change threatens the Right to Health..6

Climate Change threatens the Right to Water in India 6

Climate Change threatens the Right to Adequate Housing in India 7

Climate Change threatens Minority Rights in India 8

Conclusion 9

'Human rights law is relevant because climate change causes human rights violations. But a human rights lens can also be helpful in approaching and managing climate change.' – Mary Robinson, President, Realising Rights[1]

> The movement of people in response to climate-related events will compromise their Human rights. Human Right laws will be playing an important role where there is violation of Human rights related to the impacts of climate change[2]. Under the 1945 Charter of the United Nations[3], the 1948 Universal Declaration of Human Rights[4], and other international human rights instruments, states have a specific duty to prevent the violation of human rights[5]. under Article 2 of the International Covenant on Economic, Social and Cultural Rights(ICESCR) states have an obligation to 'undertake steps, individually and through international assistance and cooperation to fulfil rights' and are required to use 'the maximum of its available resources to that end.[6] In order to live with dignity certain basic rights and freedoms are necessary, which all Human beings are entitled to inform as Human Rights. In this paper we will be analysing the Human rights impact of Climate Change across India.

Introduction

'Human rights are the fundamental moral claim of each person to life's essentials such as food, water, shelter, and security no matter how much or how little money or power they have'[7]. Particularly if the government needs to undertakes to resettle large numbers of people in India. Of growing concern there are serious gaps in the protection schemes provided by existing law, including the extent to which persons adversely affected by climate change can cross international borders for instance Climate migrants coming from Bangladesh towards India[8]. Various ambiguities and gaps in human rights and humanitarian law which leave many climate change victims/displaced who are forced to migrate unprotected and vulnerable

---

[1] International Council on Human Rights Policy (ICHRP) (2008) 'Climate change and human rights: a rough guide', Geneva: ICHRP.
[2] M Orellana and A Johl, Climate Change and Human Rights: A Primer (Washington D.C , Center for international Enviornmental Law(CIEL, 2011)P.3. S Humphreys(ed), Human rights and Climate Change (Cambridge:Cambridge University Press, 2009).
[3] the Preamble to the Charter provided that one of the purposes of the United Ntaions was to promote and encourage 'respect for human rights and for fundamental freedoms for all without distinction as to race, sex, language or religion'(Charter of the United Nations(adopted on 26 June 1945, entered into force on 24 October 1945).
[4] Universal Declaration of Human Rights (adopted 10 December 1948 UNGA Res 217 A(III)(UDHR).
[5] M. Orellana and A Johl, Climate Change and Human Rights: A Primer(Washinton DC, Center for International Enviornmental Law(CIEL, 2011), p1
[6] S Humphreys(ed), Climate Change and Human rights: A rough Guide(Versoix, Switzerland, International council on Human Rights Policy 2008).
[7] Kate Raworth Et Al., Oxfam, Climate wrongs and Human Rights: Putting People At the Heart of Climate Change Policy 20 (2008), available at http://www.oxfam.org/files/bp117-climate-wrongs-and-human-rights-0809.pdf
[8] Michelle T. Leighton , Climate Change and Migration: Key Issues for Legal Protection of Migrants and Displaced Persons (Background Paper for the Transatlantic Study Team on Climate Change and Migration, German Marshall Fund of the United States, 2010)

to abuse and facing challenge at both level from the place they are moving and the destination country. Although the language varies, a re-examine indicates that a number of key human rights are identified and recognised, including the right to life and liberty; freedom of expression, religion, movement and residence; respect for privacy, family and home and the right to property. The most fundamental human right that may be affected by climate change, for example through the increase in natural disasters and flooding, is *the right to life* which is protected under Article 6 of the International Covenant on Civil and Political Rights(ICCPR) and is considered a 'peremptory norm' of international law. The 2009 Report of the Office of the United Nations High Commissioner for Human Rights on the relationship between climate change and human rights the OHCHR study noted that the link between climate change and human rights and obligation on States to protect human rights from the effects of Climate change. On other hand it also noted that in order to access the human rights impact of climate change related phenomenon and of policies and measures adopted to address climate change there is need for more detailed study and data collection across regions[9]. The report further stated that "adverse effects of global warming are often projections about future impacts, whereas human rights violations are normally established after the harm has occurred"[10].

In this paper we are linking social and human impacts of climate change with specific human rights, right to life, food, health, housing and water and minority rights are under threat from the anthropogenic impacts of climate change across India. It is worth nothing that many of the rights are at risk from climate impacts fall in the category of economic, social and cultural rights are concerned.

Exiled residents from their native place or border crosser who are unable to return to their residences or not capable to resume their traditional livelihoods are usually forced to beginning to the urban centres in search of employment and a better life from rural areas in within [India] or across border. These people in general face insecurity of land tenure and shelter, with women especially vulnerable to exploitation and abusive practices. The conditions under which most of the rural-urban migrants live violate their most basic, human rights including lack of shelter, lack of secure tenure, and lack of access to basic services such as clean drinking, water, healthcare and education[11].

There are serious, far reaching human security impacts arising from such conditions.

Firstly, environmental and other rural-urban migrants tend to live in overcrowded slums, and contributing to the environmental degradation of the surrounding area.

Secondly, competition over already scarce basic resources such as clean water, electricity, etc, leads to increased social tension within the slum population and could result in outbreaks of conflict.

---

[9] United Nations Office of the High Commissioner for Human Rights, Report on the Relationship Between Climate Change and Human Rights, UN.Doc.A/HRC/10/61(Jan.15, 2009).

[10] Tully, Stephen, Like Oil and Water: A sceptical Appraisal of Climate Change and Human Rights, 15 AUST.INT'LL.J.213(2008) the author analyses human rights oriented and other litigation strategies in the context of environmental protection, and appraises the typical scheme of remedies flowing from human rights claims. While nothing the importance of the human rights paradigm in highlighting the impacts of climate change upon individuals.

[11] Sanjoy Biswas & Md. Akterul Alam Chowdhury. 2012.Climate Change Induced Displacement and Migration in Bangladesh: The Need for Rights-Based Solutions, Refugee Watch, 39 & 40, June & December.

Thirdly, the arrival of displaced persons in large numbers to a city also jeopardizes the city's ability to plan for the future, as overcrowding and overuse of existing amenities and services disrupt urban planning.

Finally, overcrowding and overpopulation of urban centres pose an incredible risk in terms of disasters such as floods and earthquakes, as well as a public health challenge. Given the current state of Bangladesh's cities, a mass exodus of environmental migrants from rural areas would no doubt be a cause for alarm. [12]

'Human rights norms are more protective of those who are displaced or who migrate within their country of origin than for those who migrate internationally. This is because governments have adopted certain baseline standards to protect the internally displaced, which govern the state's treatment of such persons in the course of natural disasters or armed conflict'[13]

Climate Change impacts are clearly visible all round the world including India. Geo-economically India is progressing, and it's coastal and indigenous population are bound to become climate migrant or climigrant, therefore climate change becomes a human right issue. Climate change would be affecting Right to Life, Right to Food, water, health, shelter environment, and most distressing effects can be visible in the exercise of Fundamental Rights enshrined under Article 21 of the Constitution of India. The Right of minority, advasis would be also under threat with this emergency of Climate Change. In India effects of Climate change are retreating Himalayan glaciers, infectious diseases, river systems and more extreme events. The longevity of heat waves across India has extended in recent years, leading to warmer temperatures at night and hotter days-this trend is set to continue[14] (Cruz et.al.2007). Now we shall be discussing climate change impacts on several fundamental rights across India on human race.

Climate change threatens the Right to Life in India

'Everyone has the right to life, liberty and security of person.' (UDHR, Article 3)

Climate change has direct implications for the right to life. In its January 2009 report on the climate change and human rights, The OHCHR states that:

'A number of observed and projected effects of climate change will pose direct and indirect threats to human lives. IPCC... project with high confidence an increase in people suffering from death, disease and injury from heat waves, floods, storms, fires and droughts. equally, climate change will affect the right to life through an increase in hunger and malnutrition and related disorders impacting on chid growth and development, cardio-respiratory morbidity and mortality related to ground level ozone. Climate change will exacerbate weather related

---

[12] Ibid.
[13] Michelle T. Leighton , Climate Change and Migration: Key Issues for Legal Protection of Migrants and Displaced Persons (Background Paper for the Transatlantic Study Team on Climate Change and Migration, German Marshall Fund of the United States, 2010).
[14] Cruz, R. V. et. al. 2007. Asia. Climate Change 2007: Impacts, Adaptation and Vulnerability. Contribution of Working Group 2th to the 4 Assessment Report of the Intergovernmental Panel on Climate Change. Parry, M. L. (eds), Cambridge University Press: Cambridge.

YOUR KNOWLEDGE HAS VALUE

- We will publish your bachelor's and master's thesis, essays and papers

- Your own eBook and book - sold worldwide in all relevant shops

- Earn money with each sale

Upload your text at www.GRIN.com
and publish for free

Bibliographic information published by the German National Library:

The German National Library lists this publication in the National Bibliography; detailed bibliographic data are available on the Internet at http://dnb.dnb.de .

This book is copyright material and must not be copied, reproduced, transferred, distributed, leased, licensed or publicly performed or used in any way except as specifically permitted in writing by the publishers, as allowed under the terms and conditions under which it was purchased or as strictly permitted by applicable copyright law. Any unauthorized distribution or use of this text may be a direct infringement of the author s and publisher s rights and those responsible may be liable in law accordingly.

Imprint:

Copyright © 2016 GRIN Verlag, Open Publishing GmbH
Print and binding: Books on Demand GmbH, Norderstedt Germany
ISBN: 9783668371200

This book at GRIN:

http://www.grin.com/en/e-book/349843/role-of-it-in-the-management-of-hospitals

disasters which already have devastating effects on people and their enjoyment of the right to life, particularly in the developing world. For example, an estimated 262 million people were affected by climate disasters annually from 2000 to 2004, of who over 98 percent live in developing countries[15]. Climate change by redrawing the maps of water availability, food security, disease prevalence population distribution and coastal boundaries has the potential to exacerbate insecurity and violent conflict on a large scale[16]. Threat to life are more immediate in some countries including India according to CNA report(2007, 2014) and it further stated that rising sea level are putting people and food supplies in vulnerable coastal regions like eastern India and could lead to new wave of refugee[17]. Some communities those living in the Arctic and in coastal regions are particularly at risk and are already starting to experience the adverse effects of climate change on their right to life[18].

Climate Change threatens the Right to Adequate food in India

'The State Parties to the present covenant, recognise the fundamental right of everyone to be free from hunger…' [19]

The ICESCR includes the right to adequate food an element of the right to an adequate standard of living[20] The CESCR has argued that right to food is fundamental to the inherent dignity of the human person and indispensable for the fulfilment of other human rights enshrined in the international Bill of Rights.

According to the CESCR, "even where a State faces severe resource constraints, weather caused by a process of economic adjustment, economic recession, climatic conditions or other factors, measures should be undertaken to ensure that the right to adequate food is especially fulfilled for vulnerable populations groups and individuals.

Climatic variations have a beyond compare impact upon poor and marginalised groups of the rural population. When rainfall patterns become more erratic and shorter, farmers are more vulnerable to many risks, including droughts, diseases and unpredictable market irregularities. According to a research analysis by Oxfam farmers, crop productivity has declined due to decreased rainfall. Sowing season has shifted due to changed monsoon seasons. And, pests and diseases have increased due to increased humidity and temperatures[21]

---

[15] United Nations Office of the High Commissioner for Human Rights, Report on the Relationship Between Climate Change and Human Rights, U.N Doc.A/HRC/10/61(Jan.15, 2009). para.70
[16] Brown, Oli and Crawford, Elec, Rising Temperature, Rising Tensions: Climate Change and the Risk of Violent Conflict in the Middle Est, international institute for sustainable Development(2009). See United Nations Environment Programme, from conflict to peace building: The Role of Natural Resources and the Environment. (2009).
[17] Davenport, Coral.2014.Climate Change Deemed Growing Security Threat by Military Researchers, http://www.nytimes.com/2014/05/14/us/politics/climate-change-deemed-growing-security-threat-by-military-researchers.html?_r=0
[18] Organization of American States, Inter-American Commission on Human Rights, Petition seeking Relief from Violations Resulting from Global Warming Caused by Acts and Omissions of the United States(2005).
[19] ICESCR, Article 11
[20] Limon, Marc, Human Rights and Climate Change:Constructing a case for Political Action, 33 HARV,ENVTL.L.REV.439(2009); Halvorssen, Anita M; Common but differentiated Commitments in the future Climate Change Regime-Amending the Kyoto Protocol to include Annex C and the Annex C Mitigation Fund, 18 Colo.J.In'I Envtl. Land Policy 247(2007).
[21] Available online at: http://asiapacific.endpoverty2015.org/

'In Jharkhand, Orissa and Chhattisgarh, severe droughts will reduce rice production by 40 percent of the total production. And, in Gujarat and Maharashtra, the rising sea levels will inundate and salinise coastal regions, thereby affecting agricultural productivity of the coastal regions'[22]

Climate Change threatens the Right to Health

'The State Parties to the present Covenant recognise the right of everyone to the enjoyment of the highest attainable standard of physical and mental health.' (ICESCR, Article 12)

The effects of climate change on human health in India is a broad topic, covering areas from extreme weather events to shifts in vector-borne diseases. India might experience innumerable human health effects because of climate change. These include infectious diseases such as *malaria, chikungunya*, and *water-borne illnesses*. Displacements due to the loss of housing, hunger, and injuries are some of the adverse outcomes to the population. 'Many other fields of health care in India could be impacted by climate change, including family practice, internal medicine, paediatrics, geriatrics, and psychiatry'[23]. A warmer climate could cause water-borne diseases to become more frequent, including cholera and diarrhoeal diseases such as *giardiasis, salmonellosis*, and *cryptosporidiosis*[24]. Currently, all of India's population is at risk for

contracting malaria except for those in the areas above 1700 m above sea surface. More than 973 million persons are exposed to vector-borne malarial parasites in India, and in 1998 an estimated 577,000 disability adjusted life years (DALYs) were lost due to malaria[25].

Climate Change threatens the Right to Water in India

The right to water is not clearly mentioned in the Constitution of India as constitutionally protected right but it has been interpreted by the Courts that the right to life includes the right to safe and sufficient water[26] In India, the Right to water has been protected as a fundamental human right by the Indian Supreme Court as part of the Right to Life guaranteed under Article 21 of the Indian constitution.

---

presscentre/whatsnew/public-hearing-highlights-climate-changeimpact-on-grassroots
[22] Lal, R., Sivakumar. M., Faiz, S., Rahman, A., and Islam, K., 2010, "Climate Change and Food Security in South Asia." Springer Dordrecht Heidelberg, New York London
[23] Hess JJ, Heilpern KL, Davis TE, Frumkin H. Climate change and emergency medicine: Impacts and opportunities. Acad Emerg Med 2009; 16 : 782-94.
[24] Hales S, Edwards SJ, Kovats RS. Impacts on health of climate extremes. In: McMichael AJ, Campbell-Lendrum DH,
Corvalan CF, Ebi KL, Githeko A, Scheraga JD, Woodward
A, editors. Climate change and human health: Risks and responses. Geneva, Switzerland: World Health Organization; 2003. p. 79-102

[25] Garg A, Dhiman RC, Bhattacharya S, Shukla PR. Development, malaria and adaptation to climate change: A case study from India. Environ Manage 2009; 43 : 779-89.
[26] Amy Hardberger "Life, Liberty and the Pursuit of Water: Evaluating Water as a Human Right and the Duties and Obligations it Creates" (2005) 4 Northwestern Journal of International Human Rights 331 at 352

The right to water[27] as an essential condition for survival, is not just a self-standing right[28] but is recognized as inextricably linked with other human rights such as the right to adequate standard of living, the right to the highest attainable standard of health and the rights to adequate housing and adequate food[29]

The stern review records that even a 1 degree Celsius rise in temperature will threaten water supplies for 50 million people and 5 degree Celsius rise in temperature will result in the disappearance of various Himalayan glaciers threatening water shortages for a quarter of China's population and hundreds of millions of Indians.[30]

The OHCHR report on climate change and human rights relies on the 2007 IPCC assessment to state: "Loss of glaciers and reductions in snow cover are projected to increase and to negatively affect water availability for more than one-sixth of the world's population supplied by melt water from mountain ranges. Weather extremes, such as drought and flooding, will also impact on water supplies[31]. Climate change will thus exacerbate existing stresses on water resources and compound the problem of access to safe drinking water, currently denied to an estimated 1.1 billion people globally and a major cause of morbidity and disease"[32].

In India, communities have accused major soft-drinks multinational corporations of using too much water in their operations, leaving households without access to water this led to conflicts between companies and communities are set to worsen as climate change severely reduces water availability in many developing countries[33]

Climate Change threatens the Right to Adequate Housing in India

Right to adequate housing is recognized human rights at the international level in various instruments like *UDHR, ICESCR, CEDAW, CRC, CERD* etc. which are very crucial for India. Indian Judiciary ruled that right to adequate housing[34] is essential part of Fundamental Rights guaranteed under Part-III specifically Article 21 and 19(1)(e) but in some case the Judiciary has also disappointed to people who need housing right[35].Climate change may

---

[27] CESCR, General Comment No.15 The Right to Water, E/C.12/2002/11(2003) Although the ICESCR does not explicitly include the right to water, the Committee decided that the right falls within "the category of guarantees essential for securing an adequate standard of living" and is "also inextricably related to the right to the highest attainable standard of health..and the rights to adequate housing and adequate food".
[28] CEDAW Article 14(2)(h), CRC Article 24(2)©, Geneva Convention relative to the Treatment of Prisoners of war Articles 20,26,29 and 46, Geneva Convention relative to the treatment of Civilian Persons in Time of war Articles 85,89 and 127.
[29] CESCR General Comment No.15, Supra note 14.
[30] Stren Review Chapter, Chapter III(How Climate change will Affect People Around the world).
[31] Climate Change 2007 - Synthesis Report, adopted at IPCC Plenary XXVII, Valencia, Spain, 12-17 November 2007(IPCC AR4 Synthesis Report),48-49.
[32] OHCHR report (2009) para.29.
[33] World Business Council for Sustainable Development (2008), 'Water and Sustainable Development: executive brief', available at: www.wbcsd.org/includes/getTarget.asp?type=d&id=ODk4Nw.
[34] [34] Pandey, Pradeep Kumar, Right to Adequate Housing in India: Human Rights Perspective (September 4, 2011). Available at SSRN: http://ssrn.com/abstract=1922129
[35] Pandey, Pradeep Kumar, Right to Adequate Housing in India: Human Rights Perspective (September 4, 2011). Available at SSRN: http://ssrn.com/abstract=1922129

impact upon the right to housing in many ways. The estimates of the number of people likely to be displaced by climate change range from 50 to 250 million by the year 2050[36]

As the OHCHR report of 2009 suggests Human rights guarantees in the context of Climate change include

(a) Adequate protection of housing from weather hazards;

(b) Access to housing away from hazardous zones;

(c) Access to shelter and disaster preparedness in cases of displacement caused by extreme weather events;

(d) protection of communities that are relocated away from hazardous zones including protection against forced evictions without appropriate forms of legal or other protection, including adequate consultation with affected persons"[37]

Climate Change threatens Minority Rights in India

Under article 27 of the International Covenant on Civil and Political Rights protects the minority's right that belong to their cultural group. This article stated that *'Persons belonging to [ethnic, religious or linguistic] minorities shall not be denied the right, in community with other members of their group, to enjoy their own culture'*. In India's severe floods of 2007, for example, the dalit community were struck hardest, because they lived in flood-prone areas in low-quality housing, and they received emergency relief last, if at all. Floods in 2007, in India, the National Campaign on Dalit Human Rights was urging greater attention to the plight of Dalits, Muslims and Adivasis in India[38] 'One of the most shocking examples of minorities' greater exposure to climate change is in India, where some 170 million people known as Dalits are physically, socially and economically excluded from the rest of society. As a result they and two other minorities, Adivasis and Muslims, were worst hit by the unusually severe monsoon floods in 2007'[39]. Many Dalits lived in shaky homes in flood-prone areas outside main villages, leaving them especially exposed received it at all, because relief workers did not realize that Dalits live outside the main villages, or because dominant groups took control of distribution or were given priority They were often last to get emergency relief [40]. Many minority and indigenous groups have a close interaction with

---

[36] Stern Review, Supra note 13 at 77 describing 250 million as a "conservative" assumption. McAdam, Jane and Ben Saul, Weathering Insecurity:Climate-induced Displacement and International Law, in Human Security and Non-Citizen Law, Policy and International Affairs(Alice Edwards and Carla Ferstman eds, 2009). Such calculations are controversial, however, in view of the difficulty in predicting the rate of sea level rise and in agreeing on how the phenomenon of climate induced displacement is to be defined.

[37] OHCHR Report(2009), para 38. In this respect the report notes further that the guiding principles on Internal Displacement(E/CN.4/1998/53/Add.2,annex) provide that, at the minimum, regardless of the circumstances, and without discrimination, competent authorities shall provide internally displaced people with and ensure safe access to… basic shelter and housing(Principle 18).

[38] Rachel Baird, "The Impact of Climate Change on Minorities and Indigenous Peoples," Minority Rights Group International Briefing, 2008, www.gsdrc.org/go/display&type=Document&id=3945.
[39] Ibid:2
[40] Ibid:3

natural resources in their livelihoods and cultures. Changing weather patterns that erode resources threaten the survival of whole cultures.[41]

Conclusion

Climate change has been characterised as a "profound denier of freedom of action and a source of disempowerment"[42].Climate Change will likely to disturb the realization of the rights to private and family life, property, means of subsistence, freedom of residence and movement. Climate change threatens their right to self determination as protected by both ICCPR and ICESCR[43]. Can India be able to provide assistance and protection to people displaced by climate change-related phenomena internally and as well as to the victim whose homes and jobs are being destroyed by prolonged drought, rising sea levels or of climate change who will be migrating towards India after crossing several borders? Can India afford to protect Human Rights of Refugees/displaced/migrants under the legal umbrella of International Human Rights laws?

Is Climate change offers anything new to the states most vulnerable to climate change like [India] *special emphasis* according to some scholars[44] it entirely depends upon the India's ability to leverage this discourse in negotiations vis-à-vis the international com-munity. This will require [India] *special emphasis* along with other developing and poor states of global south to invoke human rights discourses in new ways, since human rights have traditionally been concerned with the state-individual relationship. Respectively, asserting the right to development and growth may help India along with other developing nations of Global South to articulate their concerns about the impacts of climate change on their ability to protect their citizens' and [non-citizens] *our special emphasis* human rights[45].

---

[41] Minority Rights Group (2008) 'State of the World's Minorities 2008'.
[42] Human Development Report 2007/08, supra note 3 at 31. As the authors of Human Development Report observe, "[o]ne section of humanity-broadly the poorest 2.6 billion will have to respond to climate change forces over which they have no control, manufactured through political choices in countries where they have no voice".
[43] UDHR Article 13 ICCPR Articles 1 and 12(1), ICESCR Article 1 American Convention Article 22(1). United Nations Human Rights Committee General Comment No.12, The Right to self-determination of Peoples (1984), para.14.
[44] Mary Robinson.(2008) .Foreword, in Inter Council on Human Rights Policy, Climate Change and Human Rights:A Rough Guide, at iii (2008), available at http://www.ichrp.org/
[45] Ibid: 76.

YOUR KNOWLEDGE HAS VALUE

- We will publish your bachelor's and master's thesis, essays and papers

- Your own eBook and book - sold worldwide in all relevant shops

- Earn money with each sale

Upload your text at www.GRIN.com and publish for free